The Story of Nolan
Neuralink's First Human and the Dawn of a New Era

How One Man's Mind Became the Catalyst for a Technological Revolution

Tom K. Smith

Copyright ©Tom K. Smith, 2024.

All rights reserved. No part of this publication may be reproduced, distributed, or transmitted in any form or by any means, including photocopying, recording, or other electronic or mechanical methods, without the prior written permission of the publisher, except in the case of brief quotations embodied in critical reviews and certain other noncommercial uses permitted by copyright law.

Table of Contents

Introduction...3
Chapter 1: Nolan's Accident and Life Before Neuralink.. 6
Chapter 2: The Dawn of a New Era - Introduction to Neuralink... 13
Chapter 3: The Journey to Becoming the First Neuralink Patient..22
Chapter 4: The Surgery and Initial Impressions......33
Chapter 5: The Learning Curve and Advancements... 43
Chapter 6: Daily Life with Neuralink......................... 55
Chapter 7: The Potential and Future of Neuralink...68
Chapter 8: Addressing Criticisms and Misconceptions... 82
Chapter 9: A New Era of Possibilities.......................97
Conclusion.. 111

Introduction

In the rapidly evolving landscape of technology, the idea of merging the human brain with computers was once the stuff of science fiction. However, the development of brain-computer interface (BCI) technology has brought this concept into the realm of possibility. BCI allows for a direct communication pathway between the brain and external devices, opening up extraordinary opportunities for enhancing human abilities and treating neurological conditions. At the forefront of this revolution is Neuralink, a company dedicated to pioneering this technology to new heights. With a focus on developing implantable devices that can facilitate this interaction, Neuralink is not just pushing the boundaries of what's technologically possible but also reshaping our understanding of the human mind's capabilities.

Among the many groundbreaking advancements made by Neuralink, one story stands out as a beacon of hope and progress. Nolan, a man who faced the unimaginable challenge of severe paralysis after a tragic accident, became the first human to receive a Neuralink implant. His journey is not just a tale of personal resilience and determination but also a testament to the incredible potential of BCI technology. Through this narrative, we gain a unique perspective on how a device implanted in the brain can unlock a world of possibilities, offering a glimpse into a future where limitations are redefined.

This book aims to delve deep into Nolan's experience, unraveling the layers of his story to reveal the profound impact Neuralink has had on his life. By exploring this journey, we seek to provide a detailed account of how BCI technology has not only transformed one individual's daily existence but also holds the promise of revolutionizing the lives of countless others. The narrative will offer insight into the intricacies of this technology, shedding light on its current capabilities and the exciting potential it holds for the future. Through Nolan's eyes, readers will be guided through the complexities of living with a Neuralink implant, gaining a better understanding of both the science behind it and the human element that makes this story so compelling.

Chapter 1: Nolan's Accident and Life Before Neuralink

Nolan's life changed in an instant during what should have been a routine swim in a lake. The water, which had always been a source of enjoyment and freedom, became the backdrop for a moment that would alter his path forever. In a split second, a miscalculated movement led to the dislocation of his C4 and C5 vertebrae, leaving him paralyzed from the neck down. The severity of the injury meant that his spinal cord, the crucial conduit for signals between his brain and body, was severely compromised. This catastrophic event marked the beginning of an arduous journey filled with challenges that were both physical and emotional.

In the aftermath of the accident, Nolan faced the harsh reality of life with paralysis. Tasks that once seemed effortless were now impossible without assistance. Simple acts like feeding himself, moving from one place to another, or even adjusting his posture became monumental hurdles. He was thrust into a world where dependence on others was not just an option but a necessity. The initial period of adjustment was grueling, filled with a sense of loss and frustration as he navigated the complexities of his new reality.

Daily life required a complete overhaul. Nolan had to learn to rely on assistive devices and technology to regain some semblance of independence. One of the first tools he became acquainted with was the mouth stick, a simple yet essential device that allowed him to interact with touchscreens and press buttons by using his mouth. While this provided a measure of control, it was far from a perfect solution. His condition also brought about severe spasms, causing his body to jerk unpredictably, making it difficult to maintain the precise control needed to use the mouth stick effectively. Every day was a constant battle with these limitations, and even the most advanced assistive technologies of the time offered only partial relief.

Despite the overwhelming challenges, Nolan's resilience shone through. He confronted each obstacle with a determination to adapt and find new ways to engage with the world around him. However, the journey was far from easy. Each day was a reminder of the autonomy he had lost, and even with the support of his family and the use of assistive devices, there were limits to what he could achieve on his own. It was within this context of struggle and adaptation that the possibility of a new kind of technology—one that could bridge the gap between his mind and the physical world—began to emerge. Neuralink offered a glimmer of hope, a chance to transcend the boundaries imposed by his injury and reimagine what life could be.

In the quest for independence, assistive technologies became a crucial part of Nolan's daily life. These tools were designed to provide some level of control over his environment, helping him perform tasks that were otherwise impossible due to his paralysis. One of the most commonly used devices was the mouth stick—a simple tool with a padded end that could be manipulated by holding it in the mouth. With this, Nolan could touch the screen of a tablet or press keys on a keyboard. While the mouth stick offered a degree of interaction with digital devices, it came with significant limitations.

The most obvious drawback was the physical effort and discomfort involved. Holding the mouth stick for extended periods caused strain and fatigue, especially given Nolan's compromised physical condition. His spasms further complicated matters, often making it difficult to maintain the necessary precision for effective use. The unpredictable movements of his body could cause the mouth stick to slip or miss its target, leading to repeated attempts and mounting frustration. Tasks that should have taken seconds could stretch into minutes or longer, turning even the simplest actions into arduous endeavors.

Beyond the mouth stick, Nolan also explored eye-tracking software, a more advanced form of assistive technology. Eye-tracking aimed to provide hands-free control of digital devices by tracking the user's eye movements to move a cursor on the screen. In theory, this technology should have allowed Nolan to navigate his computer or tablet with ease. However, in practice, it proved to be unreliable and imprecise. The software often struggled to accurately follow his gaze, leading to frequent misinterpretations of his intentions. This inaccuracy not only hampered his ability to interact with devices effectively but also contributed to a growing sense of helplessness.

Moreover, the setup and calibration process for eye-tracking software were far from user-friendly. It required a controlled environment and a stable physical posture, conditions that were difficult for Nolan to maintain due to his spasms and limited mobility. Even when the system worked as intended, the experience was often slow and cumbersome. Tasks like typing a sentence or clicking a button, which most people take for granted, became time-consuming and labor-intensive processes. The promise of seamless interaction remained out of reach, highlighting the inadequacies of existing technology in meeting the needs of individuals with severe physical impairments.

These limitations underscored the gap between the potential of assistive technologies and the reality of their application. While tools like the mouth stick and eye-tracking software represented important steps forward, they fell short of providing the level of autonomy and efficiency that individuals like Nolan needed. It was clear that a more innovative solution was required—one that could bypass the physical barriers entirely and tap directly into the power of the mind. This realization set the stage for Nolan's involvement with Neuralink, a company that aimed to push the boundaries of what was possible and offer a radically new approach to restoring independence and control.

Chapter 2: The Dawn of a New Era - Introduction to Neuralink

Neuralink was founded with a bold vision: to bridge the gap between the human brain and the digital world, enhancing human capabilities and addressing neurological challenges in ways previously thought impossible. Established in 2016 by Elon Musk and a team of engineers and neuroscientists, the company set out to explore the frontiers of neuroscience and technology. The primary mission was to develop implantable brain-machine interfaces that could seamlessly integrate with the human nervous system, enabling direct communication between the brain and external devices. This ambitious endeavor aimed not just to improve the quality of life for individuals with neurological conditions but also to push the limits of human cognition and interaction with technology.

At the core of Neuralink's work is brain-computer interface (BCI) technology. BCIs are systems that establish a direct communication pathway between the brain and an external device, bypassing the usual pathways of muscle movement and sensory feedback. By decoding brain signals and translating them into commands for a computer or other devices, BCIs offer the potential to control technology using only the mind. This concept has profound implications, particularly for individuals with motor impairments or neurological disorders. It opens up possibilities for restoring lost functions, enhancing communication, and even augmenting cognitive abilities.

While the concept of BCIs is not entirely new, what sets Neuralink apart is its approach to developing this technology. Traditional BCIs have often relied on bulky, invasive systems with limited functionality and a steep learning curve. Neuralink's vision was to create a device that is minimally invasive, highly efficient, and capable of interpreting a wide range of neural signals with unprecedented precision. The company's key innovation lies in the development of ultra-thin, flexible threads that can be implanted into the brain. These threads, which are much thinner than a human hair, are designed to interface directly with neurons, capturing the electrical signals they emit.

The process of implanting these threads is also unique. Neuralink has developed a specialized robot that can perform the delicate procedure with a high degree of accuracy, reducing the risk of damage to brain tissue. This robot inserts the threads into the brain, targeting specific areas that correspond to desired functions, such as motor control or sensory perception. The implant, known as the "Link," is a small, coin-sized device that sits just beneath the scalp, housing the electronics needed to process and transmit the neural signals. This design allows for a wireless connection to external devices, facilitating real-time interaction without the need for cumbersome external hardware.

Neuralink's approach to BCI technology aims to make the interaction between the brain and computers as seamless and intuitive as possible. The goal is not just to restore lost functions but to create a system that can adapt and learn, potentially enhancing cognitive abilities over time. By developing a device that can be implanted with minimal disruption and used in a wide range of applications, Neuralink seeks to democratize access to advanced neural interfaces, making them a viable option for a broader population. This includes individuals with conditions like paralysis, as well as those looking to augment their natural capabilities in various domains.

In its pursuit of this mission, Neuralink has positioned itself at the forefront of a new era in human-technology interaction. By focusing on developing a BCI that is both powerful and user-friendly, the company hopes to revolutionize the way we connect with the digital world, offering new possibilities for independence, communication, and personal growth. Nolan's story serves as a pioneering example of how this technology can be applied in real-world scenarios, providing a glimpse into a future where the boundaries between mind and machine are increasingly blurred.

Nolan's introduction to Neuralink came somewhat unexpectedly, through a friend who had taken a deep interest in neuroscience and emerging technologies. This friend, Bane, had followed Nolan's journey closely since his accident, always on the lookout for any advancements that could offer new hope. One day, Bane came across news that Neuralink had received approval from the FDA to begin human clinical trials. The company's work with brain-computer interfaces seemed like it was straight out of a science fiction novel, but it also held the promise of real-world applications that could make a profound difference in Nolan's life.

Intrigued and excited by what he had read, Bane immediately reached out to Nolan with the news. In the middle of an ordinary day, Nolan received a call that would soon set him on a path to the extraordinary. Bane was already buzzing with excitement, his words tumbling over each other as he explained the concept of Neuralink and its groundbreaking technology. Nolan was intrigued but understandably cautious. This was the first time he had heard of Neuralink, and the idea of a brain implant was both fascinating and daunting. Nevertheless, the possibility of regaining some level of control and independence was too significant to ignore.

With Bane's encouragement, Nolan began to explore what Neuralink was all about. He delved into research about the company, its mission, and the potential of brain-computer interface technology. The more he learned, the more the idea captivated him. The prospect of controlling devices with his mind was almost unbelievable, yet here was a company actively working to make it a reality. After years of dealing with the limitations imposed by his paralysis and the shortcomings of existing assistive technologies, Neuralink offered a glimmer of hope that was unlike anything he had encountered before.

The decision to apply for the clinical trial was not one Nolan took lightly. He weighed the risks and benefits, considering not just what the technology might offer him personally but also how it could contribute to a larger understanding of brain-computer interfaces. Nolan was acutely aware that participating in such a trial was not without its challenges and uncertainties. It involved undergoing brain surgery, an invasive procedure with inherent risks. Moreover, being the first human subject meant that he would be venturing into largely uncharted territory, with no guarantee of success.

Despite these concerns, Nolan felt a growing sense of determination. He was driven by the possibility of making a meaningful contribution to the advancement of this technology, which could one day help countless others facing similar challenges. He wanted to be part of something that could redefine what was possible for individuals with paralysis and other neurological conditions. With the support of his family and a firm resolve, he decided to take the leap and apply for Neuralink's clinical trial.

The application process itself was a journey. Bane, who had been instrumental in introducing Nolan to Neuralink, helped him navigate the various stages. They filled out forms, gathered medical records, and documented Nolan's condition in detail. There were moments of humor and levity even amidst the seriousness of the situation, like when Bane mistakenly spelled Nolan's name wrong on the application. But through it all, Nolan remained focused on the goal, driven by the hope that this might be the key to unlocking a new level of interaction with the world around him.

As the days passed, anticipation built. Then came the moment when Nolan received confirmation that he had been selected for the trial. It was a mix of excitement and relief, coupled with the weight of what lay ahead. The next steps would involve preparing for the procedure and entering into a realm where science fiction was becoming science fact. Nolan was ready to embrace this opportunity, fully aware that he was stepping into a pioneering role that could change not only his life but potentially the lives of many others in the future.

Chapter 3: The Journey to Becoming the First Neuralink Patient

Applying for the Neuralink clinical trial was not a straightforward process; it required Nolan to undergo a series of evaluations and preparations. The first step involved an extensive application that asked for detailed information about his medical history, the nature of his injury, and his current physical condition. Given the pioneering nature of the trial, Neuralink needed to ensure that candidates met specific criteria to minimize potential risks and maximize the chances of successful outcomes. Nolan and his friend Bane meticulously filled out the application, providing a thorough account of his injury and the various treatments and technologies he had tried over the years.

In addition to the basic application, Nolan had to provide medical records, including detailed reports on the dislocation of his C4 and C5 vertebrae. These records were crucial for Neuralink's team to assess whether his condition made him a suitable candidate for the implant. They needed to understand the extent of his paralysis and ensure that the implant could be safely and effectively positioned to establish the desired neural connections. This process involved multiple rounds of assessments, including consultations with neurologists and neurosurgeons who would later be involved in the surgical procedure.

Neuralink also conducted a series of interviews and tests to evaluate Nolan's cognitive abilities and psychological readiness for the trial. They needed to ensure that he was not only physically suitable but also mentally prepared for the challenges that lay ahead. Participating in a trial of this nature required a significant amount of resilience and adaptability, as Nolan would be navigating uncharted territory with potential risks and unknowns. His determination and optimistic outlook shone through during these assessments, reinforcing the belief that he was an ideal candidate for this groundbreaking endeavor.

After what felt like an exhaustive selection process, Nolan finally received the news he had been hoping for—he had been chosen to be the first human participant in Neuralink's clinical trial. This was both an honor and a heavy responsibility. While he was excited about the possibilities the implant could offer, he was also acutely aware of the implications of being the first person to undergo this procedure. There was no road map to follow, and the success of the trial would depend on his ability to adapt and provide feedback to the Neuralink team.

With the selection confirmed, the next step was preparing for the surgery to implant the Neuralink device. This was not a decision made lightly. It involved detailed discussions with the surgical team about the procedure, the potential risks, and the expected outcomes. Nolan and his family were briefed on every aspect of the surgery, from how the device would be implanted to the post-operative care and monitoring that would be required. The procedure would involve inserting ultra-thin threads into specific areas of his brain, designed to interface directly with the neurons responsible for motor control and other functions. This delicate operation needed to be performed with extreme precision, and Neuralink had developed a specialized robotic system to assist with the implantation.

The days leading up to the surgery were filled with a mix of anticipation and preparation. Nolan underwent a series of pre-operative tests, including brain imaging scans to map out the exact areas where the threads would be placed. These scans provided a detailed blueprint for the surgical team, ensuring that the implant would be positioned in the most effective and safe manner. He also had to follow specific pre-surgery protocols, such as adjusting his medication regimen and adhering to dietary restrictions to minimize any risks during the procedure.

In addition to the medical preparations, there was an emotional and psychological component to getting ready for the surgery. Nolan spent time with his family, discussing their hopes and concerns for the outcome. While he was ready to take this step, he knew that it would bring about significant changes, not just for him but for those closest to him as well. His family stood by him with unwavering support, understanding the potential this technology had to transform his life. They were there every step of the way, providing the encouragement and reassurance he needed as he prepared to undergo this revolutionary procedure.

The night before the surgery, Nolan found himself reflecting on the journey that had brought him to this point. From the moment of his accident to the struggles with assistive technologies, to now being on the brink of something truly groundbreaking, it had been a journey marked by resilience and hope. As he lay in the hospital bed, he felt a mixture of nerves and excitement, knowing that the next day could be the beginning of a new chapter in his life—one where the boundaries between mind and machine would be redefined.

Nolan's journey toward becoming the first human to receive a Neuralink implant was not one he undertook alone. His family played an instrumental role in every step of the process, providing the emotional foundation and unwavering support that helped him navigate this uncharted territory. From the moment Nolan learned about the possibility of participating in Neuralink's clinical trial, he knew that this decision would affect not just him but everyone close to him. It was crucial to have open and honest conversations with his family, ensuring they were fully informed and supportive of the path he was considering.

The decision to undergo brain surgery, especially for a pioneering technology like Neuralink's implant, was fraught with unknowns and potential risks. Nolan was aware of the weight of this choice and how it could impact his family. He made it clear to them that if they had any reservations or felt uncomfortable with the idea, he would reconsider his participation. His family, however, understood the magnitude of what this opportunity represented. They recognized that this was not just about Nolan potentially regaining some measure of independence, but also about contributing to a technological breakthrough that could change the lives of countless others in the future.

In the discussions that followed, Nolan's family took the time to educate themselves about Neuralink and the implications of the clinical trial. They listened to explanations of the procedure, the science behind the technology, and the potential outcomes. They also asked probing questions to ensure they fully grasped the risks involved. These conversations were crucial in building a shared understanding of what lay ahead. Nolan's family did not just accept his decision passively; they actively engaged in the process, wanting to be fully informed partners in this journey.

Their support went beyond just understanding the technology. They were there to provide emotional strength, helping Nolan weigh the pros and cons of such a significant decision. They understood that while this was a groundbreaking opportunity, it also came with the inherent risks of being the first to undergo a new type of brain surgery. There were no guarantees of success, and the outcomes were largely unknown. Despite these uncertainties, Nolan's family stood by him, encouraging him to pursue what he believed could be a life-changing experience. They trusted his judgment and respected his desire to contribute to something that had the potential to push the boundaries of what was possible for people with paralysis.

In the lead-up to the surgery, the family's support was more evident than ever. They accompanied Nolan to medical appointments, helped him prepare for the procedure, and were present for every moment of the pre-operative process. Their presence provided a sense of normalcy and comfort in what could have been an overwhelming experience. Knowing they were there, ready to support him no matter the outcome, gave Nolan the confidence he needed to face the uncertainties of the surgery. Their encouragement and belief in his decision reinforced his resolve to move forward with the trial.

On the day of the surgery, the family's role shifted from active participants to supportive witnesses. They waited anxiously while Nolan underwent the procedure, hoping for the best but prepared to handle whatever the outcome might be. Their anxiety was mixed with hope, as they understood that this was the first step in a journey that could lead to unprecedented advancements not only for

Nolan but for others in similar situations. When the surgery was completed, and Nolan emerged from the operating room, their relief was palpable. They were there to greet him, to celebrate the successful implantation of the Neuralink device, and to begin the process of recovery and adaptation together.

Throughout Nolan's journey with Neuralink, his family remained a constant source of support and motivation. They witnessed firsthand the transformative potential of the implant as Nolan began to interact with the world in ways that had been impossible since his accident. Their unwavering belief in him and in the technology helped carry him through the challenges and triumphs of being the first human to embark on this path. In many ways, the family's role was just as pioneering as Nolan's—they were breaking new ground in understanding and adapting to a reality where the mind could interface directly with machines. Their love, dedication, and willingness to embrace the unknown were vital components of

Nolan's success and a testament to the strength of the bonds that held them together.

Chapter 4: The Surgery and Initial Impressions

The surgical procedure to implant the Neuralink device into Nolan's brain was a marvel of modern medical technology, representing years of research and development. It was a delicate operation, requiring utmost precision to ensure the threads were implanted correctly to interface with the desired neurons. Unlike traditional brain surgeries that often involve large incisions and prolonged recovery times, Neuralink's approach aimed to be minimally invasive. The team used a specialized robotic system designed to implant the ultra-thin, flexible threads into specific areas of the brain with incredible accuracy, minimizing potential damage to surrounding tissues.

On the day of the surgery, Nolan was prepped and taken to the operating room, where the procedure would take place. The process began with a small incision in the scalp to allow access to the skull. The surgical robot then carefully drilled a small hole in the skull, just large enough for the implant to fit snugly. The robot's advanced imaging and guidance system allowed it to navigate to the precise locations within the brain where the threads needed to be inserted. These threads, thinner than a human hair, were designed to detect and transmit neural signals. With the robot's guidance, they were carefully placed into the target areas, ensuring that they interfaced directly with the neurons responsible for motor control and other functions.

The entire procedure was completed in under two hours, much faster than traditional brain surgeries, thanks to the precision and efficiency of the robotic system. This swift process reduced the overall risk and allowed for a shorter and more comfortable recovery period. Once the threads were in place, the device itself, a small, coin-sized unit, was implanted beneath the scalp. The incision was then closed, leaving only a minimal scar that would be almost invisible once healed.

Nolan was closely monitored in the immediate aftermath of the surgery to ensure there were no complications. To everyone's relief and astonishment, his recovery was remarkably smooth. He experienced none of the common post-surgical complications like excessive swelling, bleeding, or infection. Within just a day of the operation, he was alert and eager to move forward with the next phase of his journey. The surgical team was equally amazed by the ease of his recovery, noting how seamlessly the procedure had gone and how quickly Nolan bounced back.

As he emerged from the effects of anesthesia, Nolan was filled with a mixture of emotions. There was a sense of anticipation, knowing that the Neuralink device was now a part of him, embedded in his brain and ready to be activated. The thought that he had taken this monumental step was both exhilarating and surreal. In the days that followed, he began to process what this meant for his future. He was not just recovering from surgery; he was standing on the threshold of a new era where the boundaries between mind and machine could be explored and expanded.

Physically, Nolan felt surprisingly well. The procedure had been so minimally invasive that he experienced only minor discomfort around the incision site, and even that was manageable with basic pain relief. The medical team advised him to take it easy for the next few days, allowing his body to fully recover before they started the process of calibrating the device. Nolan appreciated the precaution, but he was eager to begin testing the capabilities of the Neuralink implant.

Emotionally, there was a sense of relief mixed with cautious optimism. The surgery had been a success, and Nolan had crossed the first major hurdle. Now came the part that everyone was most excited about: seeing how well the device would work in practice. He was about to embark on an unprecedented journey, one that would involve not just learning to use the implant but also providing invaluable feedback to the Neuralink team. Nolan felt a profound sense of purpose. He was ready to take on the role of a pioneer, exploring the potentials of this new technology and helping to shape a future where the mind's power could be harnessed in ways never before imagined.

As he lay in his hospital bed, recovering from the surgery, Nolan couldn't help but feel a surge of hope. For the first time in years, he was looking forward to the possibilities that the next day might bring. The accident that had once defined his life now seemed like the prelude to something much greater. With Neuralink's implant in place, Nolan was on the brink of discovering a whole new way to interact with the world around him, and he was ready to embrace whatever came next.

The anticipation in the room was palpable as Nolan prepared for the first session to test the Neuralink implant. This was the moment everyone had been waiting for—the point where theory and hope would meet reality. Nolan was seated in front of a computer screen, surrounded by the Neuralink team, who had been meticulously preparing for this trial run. They had spent hours setting up the system, calibrating the implant, and ensuring everything was in place for what was about to happen. The air was charged with a mixture of excitement and anxiety; even with the technology's promise, there were no guarantees about what would happen next.

Nolan focused on the screen in front of him, the cursor sitting motionless at the center. The goal was simple yet extraordinary: he would attempt to move the cursor using only his thoughts. The team explained the process to him, encouraging him to visualize the movement in his mind. It was an exercise in pure concentration, relying on the neural connections that the implant was designed to tap into. Nolan took a deep breath, his mind zeroing in on the task. He pictured the cursor shifting to the right, willing it to move with nothing but the power of his intention.

For a moment, nothing happened. The cursor remained still, and Nolan felt a flicker of doubt. But then, almost imperceptibly at first, the cursor began to inch its way across the screen. Nolan's eyes widened in disbelief. He wasn't physically moving, yet the cursor responded as if it were an extension of his own body. It was a surreal experience, a collision of thought and action that transcended the physical limitations he had been confined to for so long. The room erupted in a mix of cheers and gasps as the team witnessed this groundbreaking moment. For Nolan, it was as if a door had opened—a door to possibilities that had once seemed out of reach.

The initial shock quickly gave way to exhilaration. Nolan's mind raced with the implications of what he had just done. He had moved the cursor simply by thinking about it, by tapping into a realm of interaction that went beyond traditional physical means. It was a level of control he hadn't experienced since his accident, a direct line from his thoughts to the digital world around him. In those first moments of control, the implant had done more than just move a cursor; it had redefined Nolan's understanding of what was possible. The neural pathways that had been dormant, the signals that had been trapped within his mind, were now finding a new way to express themselves.

As Nolan continued to practice, the movements became more fluid and precise. He experimented with different directions, speeds, and tasks, pushing the boundaries of what the implant could do. Each successful action reinforced the connection between his mind and the technology, building his confidence and expanding his sense of agency. It was a learning process for both him and the Neuralink team, who carefully observed and recorded every response, taking notes on how the system adapted to his neural signals. Nolan's feedback was invaluable, providing insights that would help refine the technology for future applications.

What struck Nolan the most during these initial sessions was the naturalness of the experience. Although the sensation of moving something with his mind was new and astonishing, it also felt strangely intuitive, as if he was rediscovering a lost ability. There was a seamlessness to the interaction that defied the complexity of the technology behind it. The implant had become an extension of his will, translating his thoughts into action with a simplicity that belied the sophistication of the process. In those moments, the boundary between his mind and the machine blurred, opening up a new dimension of interaction and possibility.

The impact of this newfound control was immediate and profound. For years, Nolan had been living in a world where his physical limitations dictated every aspect of his life. Now, with the Neuralink implant, he was beginning to reclaim some of the autonomy that had been taken from him. It was more than just the ability to move a cursor; it was the beginning of a new way of living, a future where his thoughts could once again translate into direct, purposeful action. The amazement of those first moments stayed with him, a reminder of the incredible potential of this technology and the doors it could open for others like him.

In that instant, as he moved the cursor with nothing but the power of his mind, Nolan felt the true weight of what this meant—not just for him, but for the countless others who might one day benefit from this groundbreaking technology. It was a glimpse into a future where the mind's potential was no longer limited by the physical world, a future where the seemingly impossible could become possible with just a thought.

Chapter 5: The Learning Curve and Advancements

In the days following the first successful demonstration of controlling the cursor, Nolan embarked on a series of intensive brain-computer interface (BCI) training sessions. These sessions were designed to help him gain a deeper understanding and mastery of the Neuralink implant. The early stages were a learning experience for everyone involved, with Nolan at the forefront, navigating a completely new way of interacting with technology. The initial goal was to establish a consistent and reliable link between his neural signals and the commands needed to operate digital interfaces, like navigating a computer screen or selecting specific targets.

Nolan quickly discovered that using the implant required a level of mental discipline and focus unlike anything he had experienced before. The training sessions began with simple tasks, such as moving the cursor in different directions or clicking on specific icons on the screen. These seemingly basic actions were anything but easy; they demanded a great deal of concentration and the ability to isolate specific neural signals associated with intention and movement. Nolan had to train his mind to think in a new way, learning to translate his intentions into commands that the implant could interpret.

The learning curve was steep, but Nolan approached it with the same determination that had carried him through the challenges of paralysis. Each session was a process of trial and error, with moments of success interspersed with periods of frustration. Sometimes the cursor would move erratically or fail to respond at all, requiring Nolan to recalibrate his mental approach. The Neuralink team was there to support him every step of the way, offering guidance and making real-time adjustments to the system based on his performance. They encouraged Nolan to focus not on the setbacks but on the incremental progress he was making with each attempt.

What became clear early on was that Nolan was learning to control the implant at a pace that exceeded everyone's expectations. The rapid progress he made was a testament to both his own adaptability and the potential of the Neuralink technology. Within just a few sessions, he was able to move the cursor with increasing accuracy and speed. Tasks that had initially required intense concentration began to feel more intuitive. Nolan developed a kind of mental shorthand, a way of thinking about movement that allowed him to interact with the system more fluidly. It was as if his brain was rewiring itself, establishing new pathways to communicate with the digital world.

As Nolan's proficiency grew, the complexity of the tasks was gradually increased. He moved on from simple cursor movements to more intricate activities, such as selecting multiple targets in sequence, navigating through menus, and even playing basic computer games. Each new challenge pushed the boundaries of what the implant could do, providing valuable data that the Neuralink team could use to refine the technology. Nolan's ability to adapt and improve with each session was not just a personal triumph; it was a critical component in advancing the understanding of how BCI technology could be utilized.

Throughout this process, the data collected from Nolan's sessions was invaluable to Neuralink's development. Every movement, every command, and every moment of success or failure was meticulously recorded and analyzed. This data provided insights into how the brain interacts with the implant, revealing patterns in neural activity that could be optimized for better performance. Nolan's feedback was equally important. He provided the team with firsthand accounts of what it felt like to use the implant, describing the nuances of the experience in detail. His input helped the engineers and neuroscientists at

Neuralink fine-tune the system, making adjustments to improve its responsiveness and accuracy.

Nolan's unique ability to articulate his experiences was a game-changer for the Neuralink team. He could describe not only the physical aspects of using the implant but also the cognitive strategies he developed to enhance his control. This feedback loop between user and developer was crucial in refining the technology, allowing the team to make iterative improvements that directly addressed the challenges Nolan encountered. For instance, they adjusted the sensitivity of the implant to better align with Nolan's neural signals, resulting in smoother and more precise movements. They also tweaked the software to enhance the speed of response, reducing the latency between Nolan's intention and the cursor's action on the screen.

These early training sessions were more than just a test of the technology; they were a collaborative effort to push the limits of what was possible with

BCI. Nolan's rapid progress demonstrated the potential for this technology to become a practical tool for those with physical limitations. It showed that with the right training and system refinements, the mind could indeed interface directly with machines in a way that was both effective and empowering. The data and feedback collected during these sessions laid the groundwork for future advancements, not just for Nolan but for the entire field of brain-computer interfaces.

Through these initial sessions, Nolan experienced a transformation in how he perceived his own abilities. What began as a challenging and often frustrating process evolved into a journey of discovery and empowerment. He was learning to harness the power of his mind in a new and profound way, and in doing so, he was helping to shape the future of BCI technology. His contributions went beyond his own personal gains; they represented a step forward in the quest to unlock the full potential of the human brain in partnership with cutting-edge technology.

As Nolan continued his training with the Neuralink implant, his progress began to surpass even the most optimistic expectations. What initially required immense concentration and effort soon became almost second nature to him. With each session, he pushed the boundaries of what was possible, setting new benchmarks in the process. Nolan's ability to adapt and master the use of the implant quickly evolved into a series of record-breaking achievements that captured the attention and admiration of the Neuralink team and the broader scientific community.

One of the most significant milestones in Nolan's journey was his success in target selection using the brain-computer interface. Early on, simply moving the cursor toward a target on the screen was a major accomplishment. However, Nolan's skill level increased so rapidly that he soon began performing these tasks with astonishing speed and accuracy.

During training sessions, he was tasked with selecting specific targets on the screen, a process that required precise control and coordination between his thoughts and the implant. As he honed this ability, Nolan started hitting targets at a rate that far exceeded any previous attempts recorded in the field of BCI.

In one session after another, Nolan consistently selected hundreds of targets with an accuracy and speed that left observers in awe. The numbers spoke for themselves: he had successfully selected over 9,000 targets within a remarkably short period. This was an unprecedented feat, setting a new standard for what could be achieved using a brain-computer interface. For comparison, even the most advanced animal subjects, like Neuralink's well-known test subject Pager the monkey, had only managed a fraction of this performance. Nolan was venturing into uncharted territory, demonstrating a level of control and proficiency that many had thought would take years to achieve, if at all.

But it wasn't just about selecting targets. Nolan's use of the implant expanded into more complex interactions, including executing multiple types of clicks, such as left-clicks and right-clicks, with different intensities and sequences. The implant allowed him to perform over 111,000 left-clicks and 35,000 right-clicks, numbers that were staggering in their magnitude. These actions required not only precision but also the ability to rapidly switch between different types of commands, showcasing the versatility and adaptability of the Neuralink device. Nolan's achievements were setting new records in real-time, providing empirical evidence of the implant's potential to enable intricate and sustained interactions with digital interfaces.

One of the most memorable moments during this period of record-breaking progress was Nolan's engagement with more complex tasks, like playing chess and other strategic games. These games demanded a higher level of cognitive function, requiring him to think several steps ahead and execute commands quickly and accurately. During these sessions, Nolan demonstrated not only his ability to control the implant but also the capacity to use it in dynamic and challenging environments. He was able to play chess against a Neuralink team member, winning several matches with ease, a testament to his growing mastery of the device. This not only showcased his mental acuity but also the practical applications of BCI technology in enabling individuals to engage in complex activities independently.

Beyond the numbers and records, what Nolan achieved during these sessions represented a profound shift in the understanding of BCI technology's capabilities. He had proven that it was possible to interact with digital systems at a level of speed and precision that rivaled traditional physical methods. More importantly, he had done so using only the power of his mind. These records were more than just statistics; they were milestones that illustrated the immense potential of neural interfaces to restore and enhance functions that had once seemed lost forever.

For the Neuralink team, Nolan's achievements provided invaluable data and insights. His rapid progress and ability to set new records offered a unique opportunity to study the upper limits of what could be achieved with the current version of the implant. It also provided a roadmap for future developments, highlighting areas where the technology could be refined and expanded. Nolan's record-breaking performances were not just a personal victory but a leap forward for the entire field of neuroscience and technology. They showed that with the right tools and training, the human mind could interface with machines in ways that were both sophisticated and profoundly empowering.

Nolan's journey was a testament to the power of determination and the incredible potential of emerging technologies. Through his efforts, he not only set records but also redefined what was possible for individuals with physical limitations. He became a pioneer in the truest sense, blazing a trail for others to follow and providing a glimpse into a future where the boundaries between thought and action are seamlessly integrated. His record-breaking achievements with the Neuralink implant were more than just remarkable—they were a harbinger of a new era where the mind's potential could be harnessed and expanded in ways previously unimaginable.

Chapter 6: Daily Life with Neuralink

With each passing day, the Neuralink implant transformed Nolan's life in ways that extended beyond the confines of the lab. The technology opened up new avenues of independence that had once seemed impossible. One of the most significant changes was his newfound ability to engage in activities that had been out of reach since his accident. The simple yet profound act of playing video games was one of these activities. Before Neuralink, Nolan could only attempt to play games using assistive devices like mouth sticks, which were cumbersome and prone to disruptions caused by his spasms. These limitations made gaming frustrating and often short-lived, stripping away much of the joy he once found in it.

With the Neuralink implant, however, the experience was entirely different. The implant allowed him to control the games directly with his mind, bypassing the physical constraints that had previously held him back. For the first time in years, Nolan found himself immersed in the digital world of games like "Civilization VI." He could play for hours on end, navigating complex strategies and making decisions with a level of precision that was previously unattainable. This newfound freedom was more than just a form of entertainment; it was a reclaiming of a part of his life that he had lost. Being able to play without assistance or interruptions gave Nolan a sense of normalcy and autonomy, allowing him to fully engage in an activity he loved.

Beyond gaming, Neuralink also revolutionized how Nolan interacted with digital interfaces in general. He could now navigate through menus, browse the internet, and interact with software applications with a degree of control that was both empowering and liberating. Tasks that once required the help of others or the use of cumbersome assistive technologies could now be performed independently and seamlessly. This extended to everyday activities like sending emails, browsing social media, and exploring new information online. Nolan no longer felt like a passive observer in the digital age; he was an active participant, with the ability to control and engage with technology in a way that was tailored to his needs.

However, the journey to enhanced independence was not without its challenges. Adapting to the Neuralink implant required Nolan to rethink how he approached interactions with the digital world. The initial learning curve, while manageable, involved periods of trial and error as he learned to fine-tune the way he thought about movement and control. He had to develop a new mental discipline, training his brain to send the right signals in a consistent manner. This meant hours of practice, patience, and a willingness to confront the frustrations that came with mastering a completely new form of interaction.

One of the main challenges Nolan faced was learning to maintain focus over extended periods. The implant required him to be mentally engaged in a way that was different from using traditional input methods. For instance, to move a cursor or execute a command, he had to maintain a level of concentration that could be difficult to sustain, especially during complex tasks or games. At times, his mind would wander, or he would become fatigued, leading to less precise movements. Overcoming this required Nolan to develop strategies to keep his focus sharp, such as taking breaks to prevent mental fatigue and practicing mindfulness techniques to improve his concentration.

Another challenge was adapting to the nuances of the implant's feedback. Unlike traditional controllers or input devices, the Neuralink implant did not provide the same kind of tactile or visual feedback. Nolan had to rely on a more abstract sense of control, learning to interpret the subtle shifts in the system's responsiveness. This meant that he had to become highly attuned to the signals his brain was sending, developing an internal sense of how to adjust his mental commands to achieve the desired outcome. It was a process of learning to listen to his own mind in a new way, to understand the language of thought that the implant was interpreting.

Despite these challenges, Nolan's adaptability and resilience shone through. He approached each obstacle as an opportunity to learn and improve, working closely with the Neuralink team to refine the system and enhance its usability. His feedback was instrumental in making adjustments to the implant's software, improving its responsiveness and functionality. Together, they developed protocols to help him maintain focus and optimize the implant's performance, allowing him to gradually expand his capabilities. Nolan's perseverance paid off, as he began to experience a level of independence and engagement that had previously been out of reach.

The impact of Neuralink on Nolan's daily life extended beyond the immediate practical benefits. It gave him a renewed sense of purpose and agency, allowing him to reclaim aspects of his identity that had been overshadowed by his paralysis. With the implant, he was no longer defined solely by his limitations; he was an active participant in a groundbreaking technological journey, one that held the promise of not just restoring, but enhancing human abilities. This shift in perspective was perhaps the most profound change of all, offering Nolan a new way of living that was filled with possibilities and potential.

Through his experience, Nolan became a pioneer not just in terms of technology, but in redefining what it means to live with a physical disability in the digital age. He demonstrated that with the right tools and support, independence could be reimagined and reclaimed, offering a path forward for countless others facing similar challenges. Neuralink had not just changed the way he interacted with technology; it had changed the way he interacted with life itself, providing a bridge between his thoughts and the world that was more direct and empowering than he had ever imagined.

The impact of the Neuralink implant on Nolan's life went far beyond the physical capabilities it restored. It touched something deeper, bringing about a profound change in his mental and emotional well-being. Before the implant, Nolan's days were filled with a sense of restriction and dependency. Every action, every movement, was dictated by the limitations imposed by his paralysis. The frustrations of relying on others for even the simplest tasks had taken a toll on his psyche. While he had always shown remarkable resilience, there was an underlying current of loss that accompanied the reality of living with a condition that rendered him so dependent on external help.

Gaining a new level of independence through Neuralink was like opening a door that had been shut for far too long. Suddenly, Nolan had access to a part of his life that he thought he might never experience again. The ability to control aspects of his environment, to interact with the digital world on his terms, brought about a shift in how he viewed himself and his place in the world. The implant didn't just give him the ability to play video games or navigate a computer screen; it gave him back a sense of agency that had been stripped away by his injury.

This newfound independence had an immediate and powerful effect on Nolan's mental well-being. Where there had been a sense of stagnation and helplessness, there was now a feeling of empowerment. The realization that he could once again make decisions, take actions, and engage with the world without relying on others was incredibly liberating. It reignited a spark of purpose that had been dimmed by years of frustration. Nolan found himself looking forward to each new day, eager to explore the capabilities of the implant and to discover what else he could achieve. It was as if a part of him that had been dormant was coming back to life.

Emotionally, the journey was transformative. The victories he experienced—whether it was moving a cursor for the first time or successfully completing a complex task using the implant—were more than just technological milestones; they were personal triumphs that reaffirmed his sense of self. Each success was a reminder that he was not defined by his limitations but by his ability to overcome them. This shift in mindset had a ripple effect on every aspect of his life. Nolan began to see himself not as a passive recipient of care, but as an active participant in shaping his future. This new perspective brought with it a profound sense of hope and possibility.

However, this transformation was not without its complexities. The process of adapting to a new level of independence also came with emotional challenges. Nolan had to reconcile the excitement of gaining new abilities with the reality of the years he had spent without them. There was a bittersweet element to the experience, a recognition of what he had lost and a simultaneous celebration of what he was regaining. At times, this brought up emotions that were difficult to process—a mix of joy, relief, and a touch of sorrow for the time that had passed. Nolan had to navigate these feelings, understanding that while the implant was giving him back a part of his life, it was also changing the way he saw his past and future.

One of the most significant changes was in how Nolan perceived his own value and potential. Before Neuralink, he often felt limited by his condition, his potential seemingly capped by the physical barriers he faced. The implant changed that narrative. It opened up a world where his abilities were no longer solely defined by his physical state. This shift in self-perception had a profound impact on his mental health. He started to dream again, to set goals and imagine a future that wasn't limited by his condition. This mental and emotional rejuvenation was, in many ways, as significant as the physical benefits of the implant.

Nolan's experience with Neuralink also deepened his sense of connection to others, particularly to the team working on the project and to the broader community of people with disabilities. He became more than just a participant in a clinical trial; he was a pioneer, a symbol of what was possible with the right technology and support. This role brought a sense of purpose and responsibility, as he realized that his journey could inspire and pave the way for others. It was a new form of contribution, one that extended beyond his personal experience and into the realm of advocacy and innovation.

The emotional impact of this newfound independence also had a profound effect on Nolan's relationships with his family and friends. While they had always been his support system, the dynamic shifted as Nolan gained more autonomy. His ability to engage in activities independently reduced the constant need for assistance, alleviating some of the emotional and physical burdens that came with caregiving. This allowed his interactions with his loved ones to become less about necessity and more about shared experiences and connection. They could celebrate his achievements together, experiencing the joy of his progress without the overshadowing weight of dependency.

In gaining a new level of independence through Neuralink, Nolan also gained a renewed sense of identity and purpose. The psychological and emotional effects of this transformation were profound, reshaping how he saw himself and his place in the world. It was a journey of rediscovery, where the mind's potential was unlocked not just to control technology, but to reclaim a life filled with meaning, hope, and possibility. The implant was more than just a device; it was a catalyst for change that touched every part of Nolan's being, offering him a new narrative in which he was not just surviving, but thriving.

Chapter 7: The Potential and Future of Neuralink

As Nolan continued to explore the capabilities of his Neuralink implant, his thoughts naturally turned to the future and the transformative potential this technology could hold for others. His own experience had been nothing short of life-changing, and he couldn't help but imagine how it might impact the lives of people facing similar challenges. For Nolan, the implant was not just a tool for regaining independence; it was a glimpse into a future where the limitations imposed by physical impairments could be redefined and overcome in ways that were once thought impossible.

Nolan envisioned a world where Neuralink could provide unprecedented levels of autonomy for individuals with paralysis and other neurological conditions. He saw the potential for the implant to become an integral part of a suite of technologies designed to enhance quality of life. For quadriplegics like himself, this could mean gaining the ability to control a wide range of devices and systems using only their thoughts. From operating computers and smartphones to interacting with smart home environments, the applications seemed limitless. Nolan imagined a future where individuals could use the implant to turn on lights, adjust thermostats, and even open doors, all without needing physical assistance. This kind of integration could fundamentally change the way people with physical disabilities interact with their surroundings, offering a level of convenience and independence that was previously unimaginable.

Beyond these immediate applications, Nolan also saw the potential for Neuralink to facilitate more advanced interactions with technology. He was particularly excited about the possibilities of combining the implant with other emerging technologies, such as robotics. For instance, he envisioned a future where the Neuralink implant could be used to control robotic limbs or exoskeletons, providing individuals with a means to physically interact with the world around them. This integration could potentially allow people who are paralyzed to perform tasks that require fine motor skills, such as picking up objects or even walking with the aid of a robotic support system. The thought of being able to perform these actions again, or for the first time in years, was a powerful motivator for Nolan and underscored the revolutionary potential of this technology.

Nolan also considered the broader implications of Neuralink's technology for the medical field. He envisioned the implant being used to monitor and potentially treat neurological conditions in real-time. For instance, it could be used to detect early signs of seizures in individuals with epilepsy or to provide targeted stimulation to specific areas of the brain to alleviate symptoms of disorders like Parkinson's disease. These applications could not only improve the quality of life for individuals with these conditions but also open up new avenues for research and understanding of the brain.

In terms of communication, Nolan imagined a future where the implant could be used to facilitate direct brain-to-brain communication, allowing individuals to share thoughts or emotions without the need for spoken words. This idea, while still in the realm of speculation, held the promise of creating deeper connections between people, especially for those who are unable to speak or communicate through traditional means. It was a vision of a world where the barriers to communication could be transcended, fostering a new kind of interaction that was both intimate and profound.

Nolan's optimistic vision was closely aligned with Neuralink's own plans for the future. The company had ambitious goals for the development of the implant, focusing on improving its functionality and expanding its range of applications. One of the key areas of advancement was miniaturization and the reduction of invasiveness. Neuralink aimed to make future iterations of the device even smaller and easier to implant, with the goal of eventually making the procedure as routine and low-risk as LASIK eye surgery. By refining the technology and the implantation process, Neuralink hoped to make the device accessible to a wider range of individuals, including those who might benefit from its capabilities but are hesitant about undergoing brain surgery.

Another significant area of development was the integration of the Neuralink implant with other forms of technology, particularly in the realm of robotics and artificial intelligence. The company envisioned creating a seamless interface between the brain and robotic systems, allowing for direct control of prosthetic limbs, robotic assistants, and even entire environments. This integration could enable individuals with physical impairments to perform tasks with a level of dexterity and precision that was previously unattainable. For example, a person with a Neuralink implant could potentially control a robotic arm with the same ease and accuracy as they would their own limb, opening up new possibilities for independent living and personal achievement.

Neuralink was also exploring the potential of using the implant for neural recording and stimulation. This capability could be used to enhance cognitive functions or to treat neurological and psychiatric conditions. By providing targeted stimulation to specific areas of the brain, the implant could potentially alleviate symptoms of depression, anxiety, or chronic pain. The company was investigating how the implant could be used to map and understand complex neural networks, contributing to the broader scientific understanding of the brain and its functions.

In addition to these advancements, Neuralink was committed to improving the user interface and experience. The goal was to make the system as intuitive and user-friendly as possible, ensuring that individuals could easily learn to use the implant and integrate it into their daily lives. This included refining the software to improve responsiveness and adaptability, as well as developing training protocols to help users maximize the implant's potential. By focusing on user experience, Neuralink aimed to make the technology not just powerful, but also practical and accessible.

Nolan's vision for the future of Neuralink was one of boundless possibilities, where the mind's potential could be harnessed to overcome physical limitations and enhance human capabilities. His own journey with the implant was just the beginning, a first step into a new era where the integration of the brain and technology could redefine what it means to be independent and engaged with the world. The advancements that Neuralink was working toward promised to take this vision even further, creating a future where the boundaries between thought, action, and interaction are seamlessly connected, offering new horizons of freedom and expression for those who need it most.

The advent of brain-computer interface (BCI) technology, exemplified by Neuralink, brings with it profound implications for society. While the primary goal of such technology is to restore and enhance abilities for individuals with physical limitations, its potential reaches far beyond therapeutic applications. BCIs offer the promise of transforming how humans interact with technology, each other, and even their own minds. However, alongside these possibilities come a host of ethical and societal considerations that demand careful reflection and discourse.

One of the most immediate societal impacts of BCI technology lies in its potential to redefine the experience of living with a disability. For individuals like Nolan, who have faced the challenges of paralysis, BCIs offer a pathway to regain control and independence. The ability to interact with the world through thought alone can provide a level of autonomy that was previously unthinkable, fundamentally altering the lives of those with severe physical impairments. This technology could enable people to work, communicate, and participate in society more fully, reducing the barriers that have traditionally limited their opportunities and contributions.

However, the potential to restore lost functions also raises ethical questions about access and equity. If BCI technology becomes widely available, who will have access to it? The cost of developing and implementing such advanced technology is substantial, and there is a risk that it could become a privilege accessible only to those with the financial means to afford it. This could lead to a new form of inequality, where only a subset of the population benefits from enhanced capabilities. Ensuring equitable access to BCI technology is a critical challenge that must be addressed to prevent a societal divide between those who can afford neural enhancements and those who cannot.

Beyond therapeutic use, BCIs have the potential to enhance human abilities in ways that extend into the realm of augmentation. The possibility of using BCIs to enhance memory, learning, or even emotional regulation could lead to a new era of human development. For example, individuals could potentially use the technology to boost cognitive functions, access information directly from digital sources, or even communicate with others through direct neural connections. While these advancements are exciting, they also raise questions about what it means to be human and the ethical implications of altering the human experience so fundamentally.

The potential for cognitive enhancement brings with it concerns about personal identity and autonomy. If we can modify our thoughts, emotions, or memories through a neural interface, what does that mean for our sense of self? The idea of "hacking" the brain to alter one's cognitive state poses profound questions about agency and consent. Who decides what is an appropriate use of such technology, and how do we ensure that individuals retain control over their own minds? These questions touch on the very core of human autonomy and the nature of free will, suggesting the need for robust ethical guidelines and protections.

Another consideration is the privacy and security of neural data. BCIs like Neuralink involve the collection and interpretation of neural signals, which are, in essence, the most intimate data we possess. If these signals can be decoded to reveal thoughts, intentions, or emotional states, the implications for privacy are enormous. Protecting this data from misuse or unauthorized access becomes paramount. The prospect of having one's thoughts potentially accessible to external systems raises concerns about surveillance and control, highlighting the need for stringent regulations and ethical standards to safeguard individuals' neural privacy.

In addition to individual concerns, the societal implications of widespread BCI adoption include potential shifts in how we interact and communicate. If BCIs enable direct brain-to-brain communication, it could transform social interactions, creating a new form of connectivity that transcends traditional language and expression. While this could lead to deeper understanding and empathy between individuals, it could also challenge existing norms and expectations around communication. The idea of sharing thoughts directly could blur the lines between private and public, altering the way we relate to each other and navigate social spaces.

There are also potential implications for workforce and education. If BCIs can enhance learning and information processing, they could change how we approach education and skill development. People might be able to acquire new skills more rapidly or access knowledge in real-time, reshaping the demands and expectations of the workplace. This could lead to increased productivity and innovation, but it could also create new pressures and disparities in the labor market. Those with access to neural enhancements could have a distinct advantage, potentially widening the gap between different segments of the workforce.

The integration of BCIs into society also raises questions about regulation and governance. As this technology evolves, it will be crucial to develop frameworks that address the ethical, legal, and social implications of neural interfacing. This includes creating standards for safety and efficacy, ensuring informed consent for users, and establishing guidelines for data protection and privacy. Governments, scientists, and ethicists will need to work together to navigate the complex landscape of BCI technology, balancing the potential benefits with the need to protect individual rights and societal values.

In considering the broader implications of BCI technology, it is clear that we are on the cusp of a new era, one that holds the promise of profound change and innovation. Neuralink's advancements in this field represent a significant step forward, but they also bring to the forefront a host of ethical and societal questions that we must address collectively. As we move forward, it will be essential to approach these challenges with care, ensuring that the development and implementation of BCIs are guided by principles that prioritize human dignity, autonomy, and equity. Only by doing so can we fully realize the potential of this technology in a way that benefits individuals and society as a whole.

Chapter 8: Addressing Criticisms and Misconceptions

Nolan's journey with Neuralink captured the public's imagination, sparking a wide range of responses from awe and admiration to skepticism and concern. His story served as a tangible example of how brain-computer interface (BCI) technology could transform lives, and it quickly became a focal point in discussions about the future of neuroscience and human-machine integration. Many people were inspired by Nolan's experience, viewing it as a testament to the incredible possibilities that lie ahead in the field of neurotechnology. They saw in his story a beacon of hope for individuals with disabilities, a glimpse into a future where physical limitations could be overcome through the power of the mind.

However, alongside this enthusiasm came a fair share of skepticism and criticism. Public perception of Neuralink and similar technologies was colored by a mix of curiosity, fear, and ethical concerns. For some, the idea of implanting a device in the brain was unsettling, raising questions about safety, privacy, and the potential long-term effects of such interventions. There were also fears about the broader implications of BCI technology, with some people worried about the possibility of mind control, loss of individuality, or the creation of a society where neural enhancements could lead to new forms of inequality and social stratification.

Critics also expressed concerns about the commercialization of neural interfaces. They questioned the motivations behind companies like Neuralink, wondering whether the drive for profit could lead to the exploitation of vulnerable individuals or the rush to market technology that has not been adequately tested. There were fears that the technology could be used for nefarious purposes, such as surveillance or manipulation, particularly if it fell into the wrong hands. These criticisms highlighted the need for transparency, ethical guidelines, and regulatory oversight in the development and deployment of BCI technology.

Misconceptions about the technology further fueled public skepticism. Some people imagined scenarios straight out of science fiction, where individuals with implants could be hacked or controlled by external forces. Others worried about the potential for unintended side effects, such as personality changes or cognitive impairments. There was also a general misunderstanding about how the technology actually worked, with some assuming it involved invasive brain surgery or the implantation of large, bulky devices. These misconceptions often stemmed from a lack of information or from sensationalized portrayals of neural interfaces in popular media.

Amidst this mix of fascination and fear, Nolan found himself in a unique position to address these concerns and provide a grounded perspective on the reality of living with a Neuralink implant. His firsthand experience offered a valuable counterpoint to the misconceptions and skepticism surrounding the technology. Having gone through the process himself, Nolan was able to speak with authority about what the implant could and could not do, and what the experience was like from a user's point of view.

From Nolan's perspective, many of the fears and criticisms were rooted in misunderstandings about the technology and its capabilities. He acknowledged that the idea of having a device implanted in the brain could be daunting, but he emphasized the care and precision involved in the surgical process. Nolan often spoke about how minimally invasive the procedure had been, with a quick recovery time and no noticeable side effects. He pointed out that the implant itself was small and discreet, a far cry from the invasive imagery that some people had in mind. For him, the risks were outweighed by the potential benefits, both in terms of his own quality of life and the broader possibilities for others.

When it came to concerns about mind control or loss of autonomy, Nolan was quick to debunk these notions. He explained that the implant did not have the capability to alter his thoughts or emotions; it simply provided a way for him to interact with technology using neural signals. The control was entirely in his hands, or rather, in his mind. He found empowerment in the ability to direct his own actions and decisions through the implant, rather than feeling as though he was being controlled by it. This distinction was crucial in helping to demystify the technology and alleviate some of the fears surrounding it.

Nolan also addressed the ethical considerations and the need for careful regulation and oversight. He understood the concerns about privacy and the potential misuse of neural data, and he was an advocate for developing safeguards to protect users' rights and autonomy. He believed that transparency and informed consent were key in ensuring that individuals understood what they were agreeing to when using such technology. Nolan emphasized the importance of ongoing research and ethical dialogue to guide the development of BCIs in a way that prioritized human dignity and well-being.

In response to criticisms about commercialization and the motives behind Neuralink, Nolan highlighted the company's commitment to advancing scientific understanding and improving lives. He saw the technology as a tool for empowerment, one that could open up new possibilities for people with disabilities and contribute to a better understanding of the brain. While he recognized the need for oversight and ethical considerations, he also believed in the potential for positive change that this technology could bring about. For Nolan, the question was not about whether Neuralink should exist, but about how to ensure it was developed and used responsibly.

Through his public appearances and discussions, Nolan played a crucial role in shaping the narrative around Neuralink and BCI technology. He provided a voice of reason and firsthand insight, helping to dispel myths and address legitimate concerns. His experience showed that while there were certainly challenges and considerations to be navigated, the potential benefits of this technology were too significant to ignore. Nolan's story became a testament to the idea that, with the right approach, BCIs could be a force for good, offering new avenues for independence, communication, and personal growth.

In the end, Nolan's perspective on the skepticism surrounding Neuralink was one of cautious optimism. He understood the fears and concerns, but he also saw the immense potential for positive impact. He believed that through education, ethical practice, and a focus on empowering individuals, the technology could be developed in a way that respects human autonomy and enhances our ability to interact with the world. His journey with Neuralink was a reminder that while the future may be uncertain, it is also full of possibilities worth exploring.

The development and implementation of brain-computer interface (BCI) technology, like that pioneered by Neuralink, come with a range of safety protocols, ethical considerations, and long-term implications that are crucial to address. The prospect of implanting devices into the human brain to facilitate direct communication with computers opens up a world of possibilities, but it also requires careful attention to the risks and ethical dilemmas involved. Ensuring the safety and well-being of individuals like Nolan, who undergo such procedures, is paramount, as is the need to navigate the broader ethical landscape that this technology presents.

One of the foremost concerns with implanting devices in the human brain is ensuring the safety of the surgical procedure. Neuralink and similar companies have worked extensively to develop minimally invasive methods for implanting their devices. The surgical process involves the use of specialized robots that can precisely insert ultra-thin, flexible threads into targeted areas of the brain. This approach is designed to minimize the risk of damaging brain tissue and to ensure that the procedure is as safe and efficient as possible. By reducing the invasiveness of the surgery, the goal is to decrease recovery time and limit potential complications such as infection, bleeding, or swelling.

Beyond the surgery itself, long-term safety is a critical concern. Neuralink has put in place rigorous protocols to monitor the implant's performance and its interaction with the brain over time. Regular follow-up assessments are essential to ensure that the device continues to function correctly and does not cause any adverse effects. This includes monitoring for any signs of infection, inflammation, or tissue response that could indicate a problem with the implant. Additionally, the technology must be designed to withstand the environment of the human body, ensuring that it remains stable and effective over extended periods.

Another key aspect of safety is the security of the data collected by the implant. Since BCIs like Neuralink's device operate by reading and interpreting neural signals, there is a significant amount of personal and potentially sensitive data involved. Protecting this data from unauthorized access or misuse is a primary ethical concern. Neuralink has taken steps to incorporate robust encryption and data protection measures to safeguard the information transmitted by the implant. However, the evolving nature of cybersecurity threats means that continuous vigilance and updates are necessary to maintain the integrity and privacy of neural data.

Ethically, the use of BCIs raises profound questions about consent, autonomy, and the potential for unintended consequences. Informed consent is a fundamental principle in the ethical use of any medical or technological intervention. Individuals considering a Neuralink implant must be provided with comprehensive information about the procedure, its risks, benefits, and potential long-term implications. This includes a clear understanding of what the technology can and cannot do, as well as the potential for unforeseen effects. Ensuring that individuals have the capacity to make an informed decision is essential, particularly when dealing with a technology as complex and potentially life-altering as a brain implant.

The question of autonomy extends beyond the initial consent to include the ongoing use and control of the implant. Users must retain the ability to manage how the device interacts with their neural activity and have the option to deactivate or remove it if they so choose. This respect for personal autonomy is critical to ensure that individuals maintain agency over their own minds and bodies. Furthermore, there must be safeguards in place to prevent any form of coercion or pressure to use such technology, particularly in contexts where it could be seen as advantageous, such as in competitive or professional settings.

The potential for enhancement and augmentation through BCIs also raises ethical concerns about fairness and equality. If neural implants become capable of enhancing cognitive functions or providing competitive advantages in certain areas, this could lead to new forms of inequality and societal division. Access to such technology could be limited by socioeconomic factors, creating a disparity between those who can afford to augment their abilities and those who cannot. Addressing these ethical concerns requires a commitment to developing policies and frameworks that promote equitable access and prevent the misuse of BCI technology for exploitative or discriminatory purposes.

Long-term implications of implanting such devices in humans extend into the realm of societal and cultural impact. As BCIs become more integrated into daily life, they could fundamentally change how we interact with technology and each other. This raises questions about the potential for dependency on such devices and the psychological effects of interfacing with technology at such an intimate level. There is a need for ongoing research to understand how long-term use of neural implants might affect cognitive processes, mental health, and overall well-being. Ethical guidelines must evolve alongside the technology to ensure that individuals are not only protected but also supported in navigating the complexities of living with a brain-computer interface.

The possibility of using BCIs for purposes beyond therapeutic or assistive applications, such as for enhancement or surveillance, presents another layer of ethical complexity. The potential for misuse of the technology by individuals, corporations, or governments is a concern that cannot be ignored. For instance, there could be scenarios where individuals are pressured into using BCIs for monitoring or productivity purposes, leading to a loss of privacy and personal freedom. Developing ethical frameworks and regulatory measures to prevent such abuses is crucial in safeguarding the rights and dignity of individuals.

In conclusion, the implantation of BCIs like Neuralink's device represents a remarkable leap forward in technology and neuroscience, offering unprecedented opportunities for enhancing human abilities and improving quality of life. However, this progress must be accompanied by a rigorous commitment to safety, ethical responsibility, and respect for individual autonomy. By addressing the potential risks and ethical dilemmas head-on, we can work towards a future where BCIs are used responsibly and equitably, maximizing their benefits while minimizing harm. This requires not only technological innovation but also a collaborative effort to develop the policies, standards, and societal norms that will guide the integration of these devices into the fabric of human life.

Chapter 9: A New Era of Possibilities

Nolan's journey as the first human participant in Neuralink's clinical trials has been more than a personal exploration; it has significantly contributed to the advancement of brain-computer interface (BCI) technology. His involvement has offered invaluable insights into how the implant functions in a real-world context, providing the Neuralink team with crucial data that has helped refine and improve the system. By serving as a bridge between the theoretical and practical applications of this technology, Nolan has played a pivotal role in shaping the future of BCIs.

One of the key contributions Nolan made was his ability to offer detailed feedback on his experiences with the implant. As someone who was living with the device on a daily basis, he provided a unique perspective on its usability, responsiveness, and impact on his life. His observations and suggestions helped the Neuralink team understand how the device interacted with his neural activity, leading to adjustments and enhancements in both the hardware and software. For instance, through his input, they were able to fine-tune the system to better interpret his neural signals, improving its accuracy and efficiency. Nolan's feedback was essential in bridging the gap between the lab environment and real-life application, ensuring that the technology was not just effective but also user-friendly.

Nolan's participation also contributed to the gathering of extensive data on neural interactions, which is vital for the ongoing development of BCI technology. By consistently using the implant and engaging in various tasks, he generated a wealth of information about how the brain communicates with external devices. This data has been instrumental in understanding the nuances of neural signal processing and the challenges of translating those signals into meaningful actions.

Nolan's progress in mastering the implant provided empirical evidence of the potential for BCIs to facilitate complex interactions with digital interfaces. His achievements in setting records for target selection and control demonstrated the feasibility of using neural signals to perform tasks that were once considered out of reach for individuals with severe physical impairments.

Beyond the technical contributions, Nolan's journey has served as a powerful narrative for the potential of BCI technology to change lives. His story has offered a concrete example of how Neuralink's implant can restore a sense of autonomy and open up new possibilities for individuals with disabilities. For people living with conditions like paralysis, Nolan's experience provides hope and a glimpse into a future where physical limitations do not have to define one's capabilities or quality of life. By sharing his journey, Nolan has helped raise awareness about the transformative potential of BCIs, inspiring both the scientific community and the public to consider the broader implications of this technology.

The impact of BCI technology, as evidenced by Nolan's experience, extends far beyond individual cases. It has the potential to revolutionize the lives of many individuals with disabilities, offering new levels of independence and interaction that were previously unattainable. For people with motor impairments, BCIs like Neuralink's implant could provide a means of controlling devices and environments using only their thoughts. This could range from simple tasks, such as operating a computer or smartphone, to more complex actions like controlling a wheelchair or navigating a smart home. By bypassing the need for physical movement, BCIs can empower individuals to interact with the world in a way that is tailored to their abilities, enhancing their independence and quality of life.

For those who have lost the ability to communicate verbally, BCIs could offer a new mode of interaction. The technology has the potential to translate neural signals associated with speech or intention into text or digital commands, enabling individuals to express themselves and communicate with others. This could be life-changing for people with conditions like amyotrophic lateral sclerosis (ALS) or locked-in syndrome, providing a means of reconnecting with their surroundings and the people they care about. The ability to convey thoughts directly through a neural interface could bridge the gap between the mind and the external world, fostering a deeper sense of connection and engagement.

In addition to restoring lost functions, BCIs hold the promise of enhancing human capabilities in ways that could redefine what it means to live with a disability. For example, integrating BCIs with other technologies like robotics could enable individuals to control robotic limbs or exoskeletons, providing new forms of mobility and interaction. This could allow people with paralysis to perform tasks that require fine motor skills or even to walk with the assistance of a robotic support system. The combination of neural control and advanced robotics could open up a world of possibilities for physical activity and interaction, transforming the way individuals with disabilities navigate their environments.

The ripple effect of these advancements could extend into various aspects of daily life, including education, employment, and social interaction. With the aid of BCIs, individuals with disabilities could participate more fully in educational programs and professional activities, using the technology to access information, communicate with colleagues, and perform tasks that were previously inaccessible. This could lead to greater inclusion and representation in various fields, as well as the breakdown of barriers that have traditionally limited opportunities for people with disabilities. By enabling individuals to engage with the world on their own terms, BCIs have the potential to foster a more inclusive society where abilities are not defined by physical limitations.

Nolan's contributions to Neuralink and the broader field of BCI technology exemplify the profound impact that such advancements can have on individual lives and society as a whole. His journey has not only demonstrated the feasibility and potential of using neural interfaces to enhance independence and interaction, but it has also sparked a broader conversation about the role of technology in redefining human capabilities. As the development of BCIs continues to evolve, Nolan's story will stand as a testament to the power of innovation and the enduring human spirit to overcome challenges and embrace new possibilities.

The future of brain-computer interface (BCI) technology is a vast and largely uncharted territory, filled with exciting possibilities that could reshape the way we understand and interact with the world. As advancements continue to accelerate, the potential applications of BCIs extend far beyond their current uses in assisting individuals with disabilities. This technology holds the promise of transforming numerous aspects of human life, from healthcare and communication to education, entertainment, and even the nature of human cognition itself.

One of the most promising frontiers for BCI technology lies in the field of medicine, where it could revolutionize the diagnosis and treatment of neurological disorders. By providing a direct interface with the brain, BCIs could enable real-time monitoring of neural activity, offering unprecedented insights into conditions like epilepsy, Parkinson's disease, and multiple sclerosis. This could lead to earlier and more accurate diagnoses, as well as the development of personalized treatment plans that use targeted neural stimulation to alleviate symptoms or slow the progression of these diseases. Furthermore, BCIs could be used to facilitate neural rehabilitation for individuals recovering from strokes or traumatic brain injuries, helping to restore lost functions through neuroplasticity and targeted training.

Another exciting possibility is the potential for BCIs to enhance cognitive abilities, opening up new realms of learning and creativity. In the future, neural interfaces could be used to directly augment memory, allowing individuals to store and retrieve information with greater ease and accuracy. This could have profound implications for education and professional development, enabling people to acquire new skills and knowledge at an accelerated pace. Imagine a world where learning a new language or mastering a complex subject could be facilitated through direct neural input, bypassing traditional learning methods and enhancing cognitive efficiency.

Beyond individual cognitive enhancement, BCIs could also enable new forms of communication and collaboration. The idea of brain-to-brain communication, where individuals can share thoughts or emotions directly through neural interfaces, is no longer purely speculative. Researchers are already exploring ways to facilitate direct neural connections between individuals, potentially creating a new form of telepathy. This could lead to a deeper level of understanding and empathy in human interactions, as well as new possibilities for creative collaboration. For example, artists, musicians, and writers could use BCIs to share and develop ideas in ways that transcend the limitations of verbal or written language, creating collaborative works that are truly co-created on a neural level.

In the realm of virtual and augmented reality, BCIs could provide a more immersive and interactive experience by directly linking the brain with digital environments. This integration could allow users to control and interact with virtual worlds using their thoughts alone, creating a seamless interface between the mind and digital media. Such capabilities could revolutionize the gaming and entertainment industries, offering new forms of experiential storytelling and interactive design. Moreover, the use of BCIs in virtual environments could have therapeutic applications, such as creating virtual spaces for mental health treatment, rehabilitation, or social interaction for individuals who are otherwise isolated.

One of the more speculative but intriguing possibilities for BCI technology is its role in the future evolution of human cognition and society. As neural interfaces become more advanced and integrated into daily life, they could potentially change the way we think, learn, and interact with the world. This raises questions about the nature of human intelligence and the potential for a new form of collective consciousness. If BCIs enable us to connect our minds with each other and with vast networks of information, we might experience a form of shared cognition or collective intelligence that enhances our individual and societal capabilities.

However, with these future horizons come important considerations and challenges. The prospect of enhancing cognitive abilities and creating new forms of communication raises ethical questions about identity, autonomy, and the potential for inequality. There is a need for careful deliberation on how to implement and regulate such technologies to ensure they are used responsibly and equitably. As we explore the possibilities of BCIs, it will be crucial to establish ethical frameworks that protect individual rights and prevent misuse, such as the potential for coercion, surveillance, or the commodification of neural data.

Another area for exploration is the integration of BCIs with other emerging technologies, such as artificial intelligence (AI) and robotics. By combining neural interfaces with AI, we could develop systems that adapt to and learn from the user's neural activity, creating personalized and intelligent interactions. This could enhance the functionality of assistive devices, making them more responsive and intuitive. In robotics, BCIs could be used to control robotic limbs or devices with high precision, providing new levels of assistance and independence for individuals with physical impairments.

The convergence of BCIs with other technologies also opens up possibilities for space exploration and the development of advanced human-machine interfaces. In the context of long-duration space missions, for instance, BCIs could be used to monitor and maintain astronauts' cognitive and psychological well-being, as well as to control spacecraft systems in environments where traditional input methods are impractical. The ability to interface directly with technology could be a critical factor in enabling humans to thrive in space and other challenging environments.

As we look toward the future, it is clear that BCI technology has the potential to expand the boundaries of human experience in ways that are both exciting and profound. From enhancing individual abilities to creating new forms of social and cognitive interaction, the possibilities are vast and varied. However, realizing this potential will require a thoughtful and ethical approach, ensuring that the technology is developed and implemented in ways that respect human dignity, promote inclusivity, and enhance the collective well-being of society.

Nolan's journey with Neuralink has provided a glimpse into what is possible when we harness the power of the mind to interface with technology. As we stand on the threshold of a new era in human-technology interaction, the future of BCI holds the promise of not only changing lives but also redefining what it means to be human. It is a future filled with possibilities that challenge us to imagine, innovate, and navigate with care the uncharted territories that lie ahead.

Conclusion

Nolan's journey with Neuralink stands as a remarkable testament to the power of human resilience and the boundless potential of technological innovation. From the moment he became the first human to receive a Neuralink implant, Nolan embarked on a path that would not only change his own life but also mark the beginning of a new era in technology and human capability. His experience with the brain-computer interface (BCI) has illuminated the incredible possibilities that arise when the human mind and machine work in harmony. Through his courage and determination, Nolan has helped to push the boundaries of what we thought was possible, offering a glimpse into a future where physical limitations can be transcended through the power of the mind.

Nolan's legacy is one of hope and pioneering spirit. His contributions to Neuralink have been invaluable in advancing the technology, providing insights that have helped to refine and enhance the system. More than that, he has become a symbol of the potential impact of BCI technology on the lives of individuals with disabilities, demonstrating how such innovations can restore independence and open up new opportunities for interaction and engagement with the world. His story has inspired countless others, showing that even in the face of seemingly insurmountable challenges, there is always room for progress and transformation. Nolan's journey is a reminder that the quest for knowledge and improvement is a fundamental part of the human experience, one that drives us to explore new frontiers and to continuously redefine our capabilities.

The road ahead for Neuralink and BCI technology is filled with promise. The advancements made so far are just the beginning of what could become a revolution in how we interact with technology and each other. As Neuralink continues to evolve, the potential applications of this technology are expanding, from therapeutic uses for individuals with neurological conditions to enhancements that could change the way we think, learn, and communicate. The integration of BCIs with other emerging technologies, such as artificial intelligence and robotics, opens up new possibilities for augmenting human abilities and creating more immersive and intuitive interfaces. As we move forward, the focus will be on refining the technology, ensuring its safety and accessibility, and exploring its implications for society.

However, this journey is not without its challenges. The ethical considerations, safety protocols, and long-term implications of implanting such devices in humans must be carefully navigated. It is essential to approach the development and deployment of BCIs with a commitment to transparency, equity, and respect for individual autonomy. The future of BCI technology holds great potential, but it also requires us to engage in thoughtful and informed discussions about its impact on human lives and society as a whole. By addressing these challenges with diligence and care, we can work towards a future where the benefits of this technology are realized in a way that enhances human dignity and well-being.

As we stand at the dawn of this new era, it is an invitation to curiosity and exploration. Nolan's story is a call to all of us to continue learning about the transformative power of technology on human lives. It challenges us to look beyond the immediate possibilities and to imagine a future where the integration of mind and machine can lead to unprecedented levels of understanding, creativity, and interaction. This journey is a collective one, requiring the collaboration of scientists, engineers, ethicists, and the broader public to shape a future that is inclusive, ethical, and empowering.

The possibilities of BCI technology are as vast as the human imagination, and we are only just beginning to scratch the surface. Whether it is restoring lost abilities, enhancing cognitive functions, or creating new forms of communication and interaction, the potential of this technology is limited only by our willingness to explore and innovate. Nolan's legacy will continue to inspire this journey, serving as a beacon for what can be achieved when we harness the power of the mind and the potential of technology.

In the end, Nolan's experience with Neuralink is more than just a story about one man and a groundbreaking piece of technology. It is a reflection of the human drive to push beyond our limitations, to seek out new ways of understanding and engaging with the world. It is a reminder that the intersection of technology and humanity is a space of incredible potential, one that offers the promise of transformation and the opportunity to redefine what it means to live, connect, and thrive. As we look to the future, may we carry with us the spirit of curiosity and the desire to explore the uncharted, knowing that each step forward brings us closer to realizing the full potential of the human mind.

www.ingramcontent.com/pod-product-compliance
Lightning Source LLC
Chambersburg PA
CBHW050314230526
45471CB00005B/2186